Reed New Zea

Rare Birds

of New Zealand

Reed New Zealand Nature Series

Rare Birds
of New Zealand

Geoff Moon

REED

Established in 1907, Reed Publishing (NZ) Ltd
is New Zealand's largest book publisher, with over 300 titles in print.

For details on all these books visit our website:
www.reed.co.nz

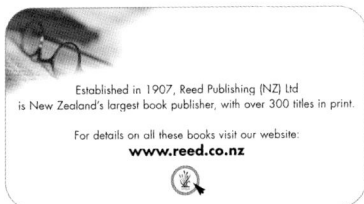

Published by Reed Books, a division of Reed Publishing (NZ) Ltd,
39 Rawene Rd, Birkenhead, Auckland. Associated companies, branches
and representatives throughout the world.

ISBN 0 7900 0371 6
First published 2000

Front cover: North Island Saddleback
Title page: Female Stitchbird

Printed in China

Contents

Introduction

This volume describes and illustrates the less common and rare birds that inhabit mainland New Zealand and its offshore islands. It does not include the species found in the Chatham Islands or outlying islands, where many rare endemic birds survive.

Some of the species described, although not often seen on the New Zealand mainland, may be reasonably common on or around outlying islands, or in other countries. For example, on the mainland the Royal Albatross is only found at Taiaroa Heads on the Otago Peninsula and surrounding seas. Yet more than 6500 pairs of this bird also nest on some outlying islands of the Chathams. Similarly, some of the migrant waders, such as the Red-necked Stint and Tattlers, are seen in very small numbers in New Zealand, yet are common in some other countries.

Included in this volume are six species that are also described in *Common Birds in New Zealand* Volumes 1 and 2. The Australasian Bittern, Reef Heron and Yellowhead are all species that have shown a decline in numbers during recent years. The New Zealand Falcon is a threatened endemic species. It is widespread but rare, and is seldom seen north of the central volcanic plateau. Although a common cosmopolitan species, the White Heron is protected in New Zealand, with an estimated total population of less than 200 birds; while the Sulphur-crested Cockatoo is rarely seen in the wild.

Taxonomists group birds into specific sections, according to their anatomy. Closely related species are listed as a genus, while related genera are grouped in families. These families are, in turn, grouped to form an order. The first part of a bird's scientific name is the genus. The second name refers to the species, and the third name, where applicable, refers to the subspecies. Thus, when we look up the Northern Royal Albatross, the family name is Diomedidae, which includes all albatrosses and mollymawks. The genus is *Diomedia*, the species is *epomophora*, and the subspecies, which distinguishes it from the Southern Royal Albatross, is *sanfordi*.

This book is intended to be used as a visual guide to identification, with scientific language kept to a minimum. The various bird species are arranged in the same order as they appear in the *Checklist of the Birds of New Zealand*, 3rd edition (Random Century, 1990), compiled by the

Checklist Committee (E.G. Turbott, Convenor) of the Ornithological Society of New Zealand Inc.

The size of each bird is given in centimetres, and represents its length from bill tip to tail tip and, in a few instances, the length of the legs extending beyond the tail. These measurements are the same as the measurements given in *Collins Guide to the Birds of New Zealand*.

Species are divided into four categories:

• *Endemic* — originating in New Zealand and confined solely to the New Zealand region, e.g., Kiwi.

• *Native* — naturally occurring in New Zealand, but also found elsewhere in the world, e.g., Blue Penguin, which also occur in

A rare and beautiful White Heron (Kotuku) with a White-faced Heron.

Australia. 'Native' includes species that are self-introduced from other countries, such as the Black-fronted Dotterel, an Australian bird which introduced itself to New Zealand.

- *Introduced* — introduced by human agency, e.g., Blackbird, Cirl Bunting.
- *Migrant* — regularly migrating to New Zealand, e.g., migrant waders such as the Bar-tailed Godwit, which arrive in their non-breeding season; and cuckoos which migrate to breed in New Zealand.

Cross-references to species covered in *Common Birds in New Zealand* Volumes 1 and 2 are shown by the abbreviation *CB1* or *CB2*, followed by the species' entry number.

All native and endemic species are protected, although some, such as the Pukeko, may be hunted in a specified open season.

I would like to acknowledge the kind assistance of many friends who have helped me with advice, or have provided opportunities to obtain photographs.

For information on all the birds inhabiting mainland New Zealand and offshore islands, refer to *The Reed Field Guide to New Zealand Birds* by Geoff Moon, published by Reed Publishing (NZ) Ltd.

Geoff Moon

Orders and Families represented in this volume

Order Apterygiformes: Kiwis
Family Apterygidae: Kiwis

Order Podicipediformes: Grebes
Family Podicipediae: Grebes

Order Procellariiformes: Tube-nosed birds
Family Diomedeidae: Albatrosses and Mollymawks

Order Sphenisciformes: Penguins
Family Spheniscidae: Penguins

Order Pelecaniformes: Gannets and Cormorants
Family Phalacrocoracidae: Cormorants and Shags

Order Ciconiiformes: Herons, Bitterns and Egrets
Family Ardeidae: Herons and Bitterns

Order Anseriformes: Swans, Geese and Ducks
Family Anatidae: Swans, Geese and Ducks

Order Falconiformes: Diurnal Birds of Prey
Family Falconidae: Falcons

Order Gruiformes: Rails
Family Rallidae: Rails and Coots

Order Charadriiformes: Waders, Gulls and Terns
Family Recurvirostridae: Stilts
Family Charadriidae: Dotterels and Plovers
Family Scolopacidae: Sandpipers, Godwits and Curlews
Family Laridae: Gulls and Terns

Order Psittaciformes: Cockatoos and Parrots
Family Psittacidae: Parrots
Family Cacatuidae: Cockatoos

Order Coraciiformes: Kingfishers
Family Alcedinidae: Kingfishers (Kookaburra)

Order Passeriformes: Perching Birds
Family Pachycephalidae: Whistlers (Yellowhead)
Family Meliphagidae: Honeyeaters (Stitchbird)
Family Emberizidae: (Cirl Bunting)
Family Callaeidae: (Kokako) (Saddleback)

Great Spotted Kiwi / Roa
Apteryx haastii

1

Family APTERYGIDAE
Genus *Apteryx*

Category
- Endemic. Endangered.

Field Characteristics
- 50 cm.
- Larger than Brown Kiwi, and has a thicker bill. Females are larger than males.
- Grey coloration with mottled and banded dark markings; chestnut tinge to feathers of back.
- Flightless; nocturnal.

Voice
- Male has a shrill, warbling whistle.
- Female call is a lower pitched, ascending whistle.

Food
- Earthworms; insects and their larvae; spiders; fallen fruit.
- Also reported taking freshwater crayfish.

Breeding
- *Time:* September to November.
- *Nest:* Usually under tree roots or in hollow fallen log. Sometimes in shallow burrow.
- *Eggs:* One egg is laid. Incubated mainly by male for 70 to 80 days. Female incubates for short shifts during night while male feeds.

Distribution & Habitat
- Great Spotted Kiwi inhabit ranges of Northwest Nelson and the Paparoa Range. There are also a few recorded sightings in beech forests further south.

Little Spotted Kiwi / Kiwi-pukupuku
2 *Apteryx owenii*

Family APTERYGIDAE
Genus *Apteryx*

Category
• Endemic. Endangered.

Characteristics
• 40 cm.
• Recognised by small size and overall grey colour. Female slightly heavier than male.
• Flightless and nocturnal.

Voice
• Male has high-pitched, trilling whistle.
• Female call lower pitched.

Food
• Earthworms; insects; spiders; fallen fruit.

Breeding
• *Time:* Early spring to late summer.
• *Nest:* Usually in burrow.
• *Eggs:* Lays one or two eggs. Incubation by male for 65 to 75 days.

Distribution & Habitat
• Little Spotted Kiwi are only common on Kapiti Island, where over 1000 birds survive.
• Recently introduced to Hen Island, Red Mercury Island, Long Island and Tiritiri Matangi.

3 Australasian Crested Grebe / Puteketeke
Podiceps cristatus australis

Family PODICIPEDIAE
Genus *Podiceps*

Category
• Native. Uncommon.

Characteristics
• 50 cm.
• Entirely aquatic. Sits low in water. Dives for its food, staying submerged for up to 50 seconds. Presumed to fly at night to other locations.

Voice
• Usually silent, but utters low-pitched groans and growls when nesting.

Food
• Dives for small fish and aquatic insects.

Breeding
• *Time:* November to February.
• *Nest:* Sticks, reeds and waterweed built on submerged willow branch or on area of compacted reed bed.
• *Eggs:* Clutch of 3 to 5 white eggs, become stained as they are covered with waterweed when bird leaves nest. Female and male share incubation for 24 to 28 days.
• *Chicks:* Able to swim soon after hatching and often ride on parents' backs.

Distribution & Habitat
• South Island only.
• Confined mainly to subalpine lakes east of Southern Alps. Also found on lowland lakes in Westland and Fiordland. Congregates on other lakes in winter when usual habitats are frozen.

Australian Crested Crebe on nest.

4 Australian Little Grebe
Tachybaptus novaehollandiae

Family PODICIPEDIAE
Genus *Tachybaptus*

Category
• Native. Uncommon.

Characteristics
• 25 cm.
• More wary than New Zealand Dabchick (*CB1*, 1). Tends to hide in rushes when approached. Conspicuous yellow 'teardrop' below eye at base of bill.

Voice
• A prolonged trill, but usually silent.

Food
• Dives for small fish, tadpoles, crustaceans and insects.

Breeding
• *Time:* September to January.
• *Nest:* Floating nest of rushes and waterweed anchored to submerged branches or rushes.
• *Eggs:* Clutch of 2 to 4 white eggs, which soon become stained. Female and male share incubation for 21 to 25 days.

Distribution & Habitat
• Self-introduced to New Zealand.
• A few birds widely scattered on small lakes in North and South Islands.

Northern Royal Albatross on nest. Note the tube nostrils on the bird's bill.

Northern Royal Albatross in flight.

5 Northern Royal Albatross / Toroa
Diomedea epomophora sanfordi

Family DIOMEDEIDAE
Genus *Diomedea*

Category
• Endemic. Uncommon.

Characteristics
• 120 to 130 cm.
• This subspecies is smaller than the southern subspecies, *D. epomophora epomophora* (Southern Royal Albatross).
• Sexes are similar in size.

Voice
• Hoarse screams and bill clapping during breeding season. Silent at sea.

Food
• Mainly squid and surface fish.

Breeding
• *Time:* Late October. Breeding starts when birds are 8 or 9 years old and takes place every second year. Approximately 25 pairs nest per year.
• *Nest:* Constructed of grasses and small sods of earth.
• *Eggs:* A single white egg is incubated by each parent for spells of 3 to 7 days. Incubation period 75 to 80 days.
• *Chicks:* Chick fed by both parents and fledges when about 37 weeks old.

Distribution & Habitat
• Occurs off South Island coasts during nesting season at Taiaroa Heads, Otago Peninsula.
• Reasonably common in sub-antarctic seas.

Yellow-eyed Penguin with chick.

6 Yellow-eyed Penguin / Hoiho
Megadyptes antipodes

Family SPHENISCIDAE
Genus *Megadyptes*

Category
• Endemic. Endangered.

Characteristics
• 76 cm.
• Considered to be the world's rarest penguin.
• Sexes similar.
• Timid and wary, these birds come ashore on certain beaches late in the day to go to their nests or to roost.

Voice
• Trumpeting and trilling calls.

Food
• Fish, squid and krill.

Breeding
• *Time:* September to December.
• *Nest:* Nests among coastal scrub and flax, always hidden from other nesting pairs.
• *Eggs:* Two eggs are very pale greenish-blue when first laid. Colour fades and eggs often become stained. Parents share incubation of 40 to 48 days.
• *Chicks:* Chicks fledge when about 15 weeks old.

Distribution & Habitat
• Inhabits east coast of Otago, Southland and Stewart Island. Occasionally seen in waters of Cook Strait.

New Zealand King Shag
Leucocarbo carunculatus

Family PHALACROCORACIDAE
Genus *Leucocarbo*

7

Category
- Endemic. Endangered.

Characteristics
- 76 cm.
- One of the rarest shags in the world.
- Easily recognisable, as it is the only pink-footed shag inhabiting the Cook Strait region.

Voice
- Silent except for low guttural croaks when nesting.

Food
- Dives from surface to catch bottom dwelling fish, such as flatfish.

Breeding
- *Time:* Usually from March to October, but can be very variable.
- *Nest:* Nest of sticks and seaweed cemented with guano.
- *Eggs:* Clutch of 2 or 3 pale blue eggs. Incubation period not known.

Distribution & Habitat
- Confined to the coastal waters on the southern side of Cook Strait.
- Roosts and nests on some of the small islands of the outer Marlborough Sounds.
- Fewer than 600 birds survive.

White Heron / Kotuku
Egretta alba modesta

Family ARDEIDAE
Genus *Egretta*

Category
• Native. Uncommon.

Characteristics
• 92 cm.
• Largest of the heron species.
• During breeding season adults grow long nuptial plumes and the normally yellow bill turns black.
• As with other herons the neck is folded back during flight.

Voice
• Guttural sounds and croaks during nesting, otherwise silent.

Food
• Eels; small fish; frogs; tadpoles; crustaceans.

Breeding
• *Time:* September to November.
• *Nest:* Nest composed of sticks.
• *Eggs:* Clutch of 2 to 5 pale blue eggs. Incubation of 24 to 26 days is shared by the parents.
• *Chicks:* Fledge at 6 weeks.

Distribution & Habitat
• A cosmopolitan species.
• Fewer than 200 birds live in New Zealand.
• Inhabits tidal estuaries and lagoons, wetlands and streams in open country.
• The only New Zealand nesting site is on the banks of the Waitangiroto River in Westland, where the birds nest in low trees and tree ferns.

White Herons at nest.

Reef Heron.

Reef Heron / Matuku moana
Egretta sacra sacra

Family ARDEIDAE
Genus *Egretta*

Category
• Native. Less common.

Characteristics
• 60 cm.
• Recognised by its overall slate-grey colour.
• Bill heavier and body more stocky than the very common White-faced Heron (*CB1*, 5).

Voice
• Guttural croaks when disturbed, or at nest.

Food
• Small fish, especially flounder; crustaceans.

Breeding
• *Time:* September to January.
• *Nest:* Built of sticks on rock ledges, in caves and rock crevices. Also occasionally in clumps of coastal flax and beneath roots of coastal trees. Some nests added to and used for many years.
• *Eggs:* Clutch of 2 or 3 turquoise-coloured eggs. Both sexes incubate eggs for 25 to 28 days.
• *Chicks:* Fledge when $5^1/_2$ to 6 weeks old.

Distribution & Habitat
• Frequents rocky shores and tidal inlets. Very rarely seen inland.
• Reef herons were once common, particularly around northern coasts. Now increasingly uncommon, possibly due to disturbance of nesting sites by recreational boating.
• A subspecies is common on many Asian coastlines and tropical Pacific islands, where a white plumage phase is common.

Little Egret
Egretta garzetta nigripes

Family ARDEIDAE
Genus *Egretta*

Category
- Native. Uncommon.

Characteristics
- 58 cm.
- Much smaller than White Heron.
- Bill and legs black. Nuptial head plumes still usually visible.
- When feeding, the bird dances and turns actively, often with wings outstretched.

Voice
- Silent.

Food
- Small fish; crustaceans; insects.

Breeding
- Has not been recorded nesting in New Zealand.

Distribution & Habitat
- Usually seen in New Zealand during autumn and winter. Probably migrates from Australia.
- Inhabits tidal estuaries and lagoons. Sometimes seen in freshwater habitats.

Nankeen Night Heron
Nycticorax caledonicus

11

Family ARDEIDAE
Genus *Nycticorax*

Category
• Native. Rare.

Characteristics
• 56 cm.
• Recognised by overall chestnut colour and squat appearance.
• Usually perches in trees during day.
• Feeds late in day and at night.

Voice
• Usually silent, but croaks at nest.

Food
• Small fish and crustaceans, particularly mud crabs.

Breeding
• Recorded nesting in Wanganui River area.

Distribution & Habitat
• Frequents tidal estuaries and mangrove swamps, as well as the Wanganui River.

Australasian Bittern / Matuku
Botaurus poiciloptilus

Family ARDEIDAE
Genus *Botaurus*

Category
- Native. Uncommon.

Characteristics
- 71 cm.
- Buff-brown plumage. Rounded wings in flight. Neck folded back when flying, as with herons.
- Usually seen singly. When disturbed, either crouches low in vegetation or points bill upwards in a frozen position, providing good camouflage when in raupo swamps. Very nervous and secretive.

Voice
- Males boom in breeding season to attract females. Also make guttural croak when disturbed to flight.

Food
- Freshwater fish, particularly eels, frogs and freshwater insects.
- Reports of taking mice and small birds.

Breeding
- *Time:* September to January.
- *Nest:* Female builds nest of raupo and vegetation.
- *Eggs:* Female lays 3 to 5 olive-brown eggs at 2-day intervals and incubates them for 24 to 26 days.
- *Chicks:* Incubation commences with laying of first eggs, so chicks vary in size after hatching. As with herons, chicks fed with regurgitated fish. Chicks wander from nest at about 14 days old and fledge and can fly at $5\frac{1}{2}$ weeks.

Distribution & Habitat
- Occurs throughout New Zealand in wetlands. Occasionally seen in mangroves.
- Bittern are becoming rare, owing to swamp drainage for agricultural use.

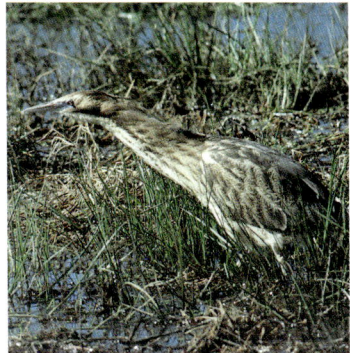

Mute Swan
Cygnus olor

Family ANATIDAE
Genus *Cygnus*

Category
• Introduced. Uncommon.

Characteristics
• 150 cm.
• Conspicuous, large white bird.
• Flies with neck extended.

Voice
• Hisses when approached near nesting area. Also utters a low, subdued whistle away from nest.

Food
• Mainly waterweed and vegetation from banks of lakes. Also willow leaves and some aquatic invertebrates.

Breeding
• *Time:* September to November.
• *Nest:* Very large nest of reeds and vegetation.
• *Eggs:* The clutch of 4 to 10 white eggs incubated by female for 35 days.
• *Chicks:* Cygnets (the young) able to swim and leave nest soon after hatching, but use nest as a base for roosting.

Distribution & Habitat
• Introduced to New Zealand in 1860s.
• Fewer than 100 birds inhabit Lake Ellesmere in South Island.
• Smaller numbers in Hawkes Bay and on ornamental lakes.

Cape Barren Goose
Cereopsis novaehollandiae

Family ANATIDAE
Genus *Cereopsis*

Category
• Introduced. Rare.

Characteristics
• 85 cm.
• Recognised by overall grey colour, long legs and erect posture.

Voice
• Repeated honking when in flight.

Food
• Grazes pasture and other vegetation. Also eats seeds.

Breeding
• *Time:* June to September.
• *Nest:* Nests near shelter. No recent recorded nesting in wild.
• *Eggs:* Lays 3 to 6 cream-coloured eggs.

Distribution & Habitat
• First introduced from Australia in 1914. Some stragglers occasionally arrive from that country.
• Inhabit open country and wetlands. Birds seen infrequently in the wild.

Blue Duck in its typical habitat.

Blue Duck / Whio
Hymenolaimus malacorhynchos

Family ANATIDAE
Genus *Hymenolaimus*

15

Category
• Endemic. Threatened.

Characteristics
• 53 cm.
• The only species of duck likely to be seen on turbulent high-country rivers.
• Well camouflaged by its overall grey colouring when perched on river rocks, though the light-coloured bill reveals the bird's presence when it moves.
• Soft, fleshy borders to bill.
• Swims and dives rapidly against current.
• Strong, direct flight.

Voice
• Male gives high-pitched whistle.
• Female gives low, rasping note when disturbed.

Food
• Insects and aquatic invertebrates taken from surface and around rocks.
• Caddis fly larvae are a favourite food.

Breeding
• *Time:* August to December.
• *Nest:* Composed of twigs and vegetation and lined with down. Usual site is in hollow log, in cavity under a rock or in thick vegetation.
• *Eggs:* A clutch of 4 to 8 cream-coloured eggs laid at two day intervals. Female alone incubates when clutch is complete, while male keeps guard nearby. Incubation period about 32 days.
• *Chicks:* Ducklings active soon after hatching and can swim and dive for food.

Distribution & Habitat
• Inhabits turbulent high-country rivers and streams in both North and South Islands, but does not occur north of central North Island.
• Recently re-introduced to some rivers on Mt Taranaki, where it was once reasonably common.

Brown Teal / Pateke
Anas aucklandica chlorotis

Family ANATIDAE
Genus *Anas*

Category
• Endemic. Threatened.

Characteristics
• 48 cm.
• Small and brown.
• Usually seen in small flocks.
• Tends to swim away when disturbed, rather than fly.
• Active at twilight and dawn.

Voice
• Male gives a hoarse rasp and female a repeated, rapid quack.
• Generally fairly silent.

Food
• Invertebrates; water insects; water plants.

Breeding
• *Time:* During many months of the year, but mainly July to January.

• *Nest:* Nest of grasses lined with down, built in dense vegetation near water.
• *Eggs:* 4 to 8 buff-coloured eggs. Female alone incubates for 27 to 30 days.

Distribution & Habitat
• Main wild population of these ducks is on Great Barrier Island, where about 1400 inhabit wetlands. Some birds are occasionally seen on the island harbours such as Fitzroy.
• Smaller populations survive on some Northland rivers and in Fiordland.
• They are successfully bred in captivity and have been liberated on offshore islands.

New Zealand Falcon (eastern race).

A New Zealand Falcon at its nest on a rock ledge.

New Zealand Falcon / Karearea
Falco novaeseelandiae

Family FALCONIDAE
Genus *Falco*

Category
• Endemic. Threatened.

Characteristics
• 43 to 46 cm.
• Rapid flight and long tail distinguish it from the Australasian Harrier, which is much larger with a more leisurely flight.
• Female falcons are markedly larger than males.
• Considered to be one variable species, with colour changes in different localities.
• The eastern race from the east of the Southern Alps is the most common of the three races. The Bush Falcon (Westland and the North Island) and the southern race (Fiordland and Stewart Island) are less common and have more rounded wings.

Voice
• Rapid repeated 'kek-kek-kek'.
• Immature birds give a subdued, high-pitched scream.

Food
• Rodents, small birds and large insects.
• Occasionally take larger prey such as rabbits, may raid poultry pens.

Breeding
• *Time:* September to December.
• *Nest:* Falcons use no nesting material. Eggs are laid in a scrape in the ground under overhanging rocks or on ledges of rocky bluffs.
• Bush-inhabiting falcons nest in tall trees, often in clumps of epiphytic perching astelia. Will also nest on the ground, sometimes in fire breaks of pine forests.
• *Eggs:* 3 to 4 buff-coloured eggs with dark russet markings. Incubation shared by parents, but mainly performed by female. Incubation period 30 to 33 days.
• *Chicks:* Female feeds chicks while male does the hunting. The smaller male chicks fledge when 32 days old, and the females when 35 days old.

Distribution & Habitat
• Bush Falcons inhabit forests and are rarely seen north of the volcanic plateau of the central North Island.
• Usually hunt along forest margins.
• The eastern race inhabits tussock high country east of the Southern Alps.

South Island Takahe
Porphyrio mantelli hochstetteri

Family RALLIDAE
Genus *Porphyrio*

Category
• Endemic. Endangered.

Characteristics
• 63 cm.
• Although possessing small wings, the Takahe is flightless.
• Much heavier than the related Pukeko (*CB1*, 23), with heavier bill, larger body and shorter, thicker legs.

Voice
• A low-pitched 'klomph'. Also a high-pitched call similar to that of the Weka (*CB2*, 11).

Food
• In the wild, feeds mainly on the succulent stems of the red tussock.
• During winter feeds on rhizomes of ferns growing in beech forest in valleys.
• On island reserves it will take a variety of grasses, fallen fruit and other vegetation.

Breeding
• *Time:* October or November.
• *Nest:* Nest built of grasses in a bower in clumps of tussock.
• *Eggs:* Two buff-coloured eggs with brown markings are incubated by both parents for 30 days.
• *Chicks:* Chicks active after hatching. They are fed small items of food by both parents for several months.

Distribution & Habitat
• Wild population of about 150 birds inhabit Murchison Range in Fiordland.
• Birds released on Tiritiri Matangi, Mana, Kapiti and Maud Islands are thriving and breeding.

Black Stilt.

Hybrid Black Stilt (foreground) with Australasian Pied Stilts.

19 Black Stilt / Kaki
Himantopus novaezelandiae

Family RECURVIROSTRIDAE
Genus *Himantopus*

Category
- Endemic. Endangered.

Characteristics
- 38 cm.
- Recognised by totally black plumage and long pink legs.
- Immature birds are partly pied, usually with white markings around breast.
- Black Stilts hybridise with the common Pied Stilts, producing offspring with variable amounts of white, mainly on face and breast.

Voice
- A repeated high-pitched 'yap-yap'.

Food
- Invertebrates, mainly water insects and earthworms.

Breeding
- *Time:* September to December.
- *Nest:* Unlike the common Pied Stilt (*CB2*, 14), Black Stilts nest in isolation and are vulnerable to predation. Nest is a scrape in shingle of a riverbank and lined with grasses.
- *Eggs:* Clutch of 2 to 4 brown eggs with black blotches. Both female and male incubate eggs for 24 to 26 days.
- *Chicks:* Chicks active soon after hatching.

Distribution & Habitat
- A highly endangered species, having been decimated owing to predation by feral cats, ferrets and stoats. Fewer than 100 birds survive in the wild.
- Population now confined to Waitaki Basin in the Mackenzie Country of the South Island.
- They inhabit riverbeds, wetlands and especially river deltas.

Migrant waders

Each year, towards the end of September and into October, thousands of wading birds arrive in New Zealand to spend the summer feeding on the mudflats of our estuaries, harbours and lagoons. These habitats provide a rich source of food comprising marine worms, molluscs and other invertebrates.

The birds, which vary from sparrow-size to domestic hen-size, migrate to New Zealand after nesting in the tundra regions of Siberia as well as Alaska.

Most of these birds belong to the family Scolopacidae, which includes the Godwits, Sandpipers and Curlews. The Bar-tailed Godwit (*CB2*, 21) and Lesser Knot (*CB2*, 20) are the two most numerous groups of waders to arrive. The third most numerous group to migrate here comprises members of the family Charadriidae. These include the Turnstone, *Arenaria interpres* (*CB2*, 19), of which 6000 arrive each summer. The rarer migrant waders to be seen in New Zealand are the Pacific Golden Plover, the Sandpipers and Tattlers.

In the New Zealand autumn the birds leave again, after spending a summer feeding and gaining considerable weight. This will sustain them on their long flight back to their nesting grounds in the arctic and subarctic tundra regions.

RIGHT: Mudflats at Miranda on the Firth of Thames provide a rich food source for migrant waders who visit New Zealand in early summer to feed on our harbours and estauries.

Pacific Golden Plover with Godwits and Turnstones.

Pacific Golden Plover
Pluvialis fulva

Family CHARADRIIDAE
Genus *Pluvialis*

Category
- Migrant wader.

Characteristics
- 25 cm.
- Similar in size to New Zealand Dotterel (*CB2*, 15), but usually appears slimmer.
- It is well camouflaged when feeding on mudflats.

Voice
- In flight gives a melodious, clear, repeated whistle.

Food
- When feeding in grasslands it mainly eats insects and earthworms.
- On mudflats it feeds on a wide range of marine invertebrates and molluscs.

Breeding
- Nests in tundra of Siberia and western Alaska.

Distribution & Habitat
- About 600 birds migrate to New Zealand each summer.
- They often feed in coastal paddocks as well as on mudflats.
- Frequently roost with Turnstones at high tide.

Pacific Golden Plover in breeding plumage.

Curlew Sandpiper with nuptial plummage.

Curlew Sandpiper
Calidris ferruginea

Family SCOLOPACIDAE
Genus *Calidris*

Category
• Migrant wader.

Characteristics
• 22 cm.
• Recognised by decurved bill (i.e., curving downwards).
• In February and March they attain a russet-coloured nuptial plumage.

Voice
• A liquid 'chirrip' when disturbed, but usually silent.

Food
• Feeds by probing deeply in mud for marine worms, molluscs and crustaceans.

Breeding
• Nests in tundra regions of Siberia.

Distribution & Habitat
• Inhabits mudflats, estuaries and sandspits.
• Often associates with Wrybill (*CB2*, 18) and Banded Dotterel (*CB2*, 16), and seen at high tide roosting with Wrybills.

Curlew Sandpiper (left) with Wrybills.

Sharp-tailed Sandpiper
Calidris accuminata

Family SCOLOPACIDAE
Genus *Calidris*

Category
• Migrant wader.

Characteristics
• 22 cm.
• Usually shows russet-coloured crown.
• In breeding plumage has boomerang-shaped markings along lower breast and flanks.

Voice
• Usually silent, but occasionally gives a double whistle when flushed.

Food
• Insects and their larvae in fresh water pools.
• Marine invertebrates and molluscs on mudflats.

Breeding
• Nests in tundra regions of Siberia.

Distribution & Habitat
• Often at shoreline salt marshes and fresh water pools.
• Also feeds in shallow water and on mudflats.

Pectoral Sandpiper
Calidris melanotos

Family SCOLOPACIDAE
Genus *Calidris*

Category
• Migrant wader.

Characteristics
• 22 cm.
• Very similar to Sharp-tailed Sandpiper. Less rufous in colour. Markings on upper breast darker, with prominent demarcation between the light-coloured underparts.

Voice
• Usually silent, but gives a rasping double call when disturbed.

Breeding
• Nests in tundra regions of northern Siberia and Alaska.

Distribution & Habitat
• Pectoral Sandpipers are less common migrants to New Zealand.
• They prefer feeding on fresh water pools and lagoons as well as salt marshes.

Red-necked Stint (centre) with Wrybills.

24 Red-necked Stint
Calidris ruficollis

Family SCOLOPACIDAE
Genus *Calidris*

Category
• Migrant wader.

Characteristics
• 15 cm.
• The smallest of the migrant waders, it is barely larger than a house sparrow and markedly smaller than the Wrybill (*CB2*, 18) and Sandpipers.
• Assumes a russet-coloured 'balaclava' plumage in late summer.

Voice
• Usually silent when in New Zealand.

Food
• Feeds on insects and small marine organisms
• Moves rapidly and pecks in a sewing machine-like action.

Breeding
• Nests in tundra regions of Siberia and western Alaska.

Distribution & Habitat
• Inhabits mudflats and estuaries, often feeding with Wrybills (*CB2*, 18) and roosting with them at high tide.

Eastern Curlew with Pied Stilts and Godwits.

Eastern Curlew
Numenius madagascariensis

Family SCOLOPACIDAE
Genus *Numenius*

Category
• Migrant wader.

Characteristics
• 63 cm.
• The largest of our migrant waders, and recognised by its large sized and very long, decurved bill.
• It is a wary bird and difficult to approach.

Voice
• A musical, flute-like warble.

Food
• Probes deeply for marine worms, crustaceans and molluscs.
• Mud crabs are a common source of food.

Breeding
• Nests in northeastern Asia and Manchuria and further north to Siberia.

Distribution & Habitat
• Small numbers migrate to New Zealand in late September. They feed on mudflats and estuaries, usually roosting separately from Godwit flocks (*CB2*, 21).

Asiatic Whimbrel
Numenius phaeopus variegatus

Family SCOLOPACIDAE
Genus *Numenius*

Category
- Migrant wader.

Characteristics
- 42 cm.
- A wary species and difficult to approach.
- Same size as the Godwit (*CB2*, 21), but recognised by its decurved bill.

Voice
- A whistling call of a repeated 'ti-ti-ti' when in flight.

Food
- Marine worms, crustaceans and particularly small crabs.

Breeding
- Nests in western Siberia and parts of northern Europe.

Distribution & Habitat
- About 150 of these migrants visit New Zealand each summer.
- They may be seen feeding in small flocks of 5 or 6 birds.
- They usually roost on the edge of Godwit flocks at high tide. Being more wary, they fly off before the Godwits.

Wandering (Alaskan) Tattler
Tringa incana

Family SCOLOPACIDAE
Genus *Tringa*

Category
• Migrant wader.

Characteristics
• 28 cm.
• The two Tattler species are difficult to differentiate in the field. The Wandering Tattler is the larger of the two.
• In breeding plumage the barred underparts extend to the under-tail coverts.

Voice
• When disturbed, the Wandering Tattler gives a rippling call of several notes, as compared with the Siberian Tattler's two note whistle.

Food
• A variety of marine invertebrates.

Breeding
• Nests in eastern Siberia and Alaska.

Distribution & Habitat
• Wandering Tattlers prefer rocky shores and rock platforms, but are also seen on shingle beaches and occasionally on mudflats.

Siberian (Grey-tailed) Tattler
Tringa brevipes

Family SCOLOPACIDAE
Genus *Tringa*

Category
- Migrant wader.

Characteristics
- 25 cm.
- Is smaller and greyer than the Wandering Tattler.
- Viewed in the hand or with a high powered telescope, it can be distinguished by a nasal groove which extends half way along its bill. In the Wandering Tattler, this nasal groove extends for two thirds the bill length.

Voice
- A high-pitched, two syllable whistle.

Food
- Marine invertebrates.

Breeding
- Nests in northern Asia and eastern Siberia.

Distribution & Habitat
- The Siberian Tattler prefers estuary and mudflat habitats, and at high tide often roosts with Wrybills (*CB2*, 18). It is a less common visitor than the Wandering Tattler.

Terek Sandpiper
29 *Tringa terek*

Family SCOLOPACIDAE
Genus *Tringa*

Category
• Migrant wader.

Characteristics
• 23 cm.
• Easily recognised by its long, upturned bill and yellowish legs. A white trailing edge to the wings is evident during flight.

Voice
• Usually silent, but a musical trill often uttered in flight.

Food
• Probes deeply for marine worms and molluscs. Also takes surface dwelling insects in brackish pools.
• A very active feeder, darting in different directions.

Breeding
• Widely scattered nesting grounds, extending from Finland to eastern Siberia.

Distribution & Habitat
• Small numbers migrate to New Zealand each summer.
• They inhabit mudflats and estuaries and are sometimes seen in brackish pools. Often roost with Wrybills (*CB2*, 18).

Terek Sandpiper, eclipse plumage.

Pair of New Zealand Fairy Terns with chicks.

New Zealand Fairy Tern (female) at nest.

Family LARIDAE
Genus *Sterna*

Category
• Endemic subspecies. Rare.

Characteristics
• 25 cm.
• The smallest and rarest of our breeding terns.
• Can be distinguished from the migrating Little Tern by its completely yellow bill and a black eye stripe ending 5 mm from base of bill. Also lacks black tip to bill.

Voice
• A high-pitched repeated 'zwit', used particularly when its nesting territory is entered.

Food
• Feeds only on live fish, which it usually captures by diving in tidal estuaries. Anchovies and small flounder are the most common species caught.

• Chicks often have difficulty in swallowing flounder fed to them by adult birds.

Breeding
• *Time:* November to early December.
• *Nest:* Nests in isolation. Nest is a scrape in the sand, usually surrounded by a patch of broken shell. This probably restricts wind-blown sand.
• *Eggs:* Two greyish-buff-coloured eggs are laid. Both female and male incubate eggs for 20 to 23 days.
• *Chicks:* Chicks wander from nest two days after hatching and can fly when 3 weeks old.

Distribution & Habitat
• Inhabits a few ocean beaches on east coast of Northland. Also some birds on South Kaipara Head.

Eastern Little Tern, eclipse plumage.

31 Eastern Little Tern
Sterna albifrons sinensis

Family LARIDAE
Genus *Sterna*

Category
• Migrant. Uncommon.

Characteristics
• 25 cm.
• Similar size to Fairy Tern, but black eye-stripe extends to base of bill. With breeding plumage it also develops black tip to its yellow bill.
• In eclipse plumage, eye-stripe does not extend to base of bill. Black outer primaries evident. Also, dark bar on carpals often evident.

Voice
• A high-pitched 'zwit'.

Food
• Shallow dives for small fish. Only live fish taken, unlike Gulls (*CB2*).

Breeding
• Nests in Australia, New Guinea and Asia.

Distribution & Habitat
• Occurs regularly in summer months throughout New Zealand. More common in northern harbours of the North Island.
• Some immature birds seen in New Zealand during winter months.

Male Kakapo.

Male Kakapo in its 'booming bowl' eating an apple (Fiordland).

Kakapo
Strigops habroptilus

Family PSITTACIDAE
Genus *Strigops*

Category
- Endemic. Highly endangered.

Characteristics
- 62 to 64 cm.
- The world's heaviest parrot. Males weigh up to 3 kg.
- Flightless, although possessing small wings.
- Nocturnal and usually solitary. In the wild, presence evident by chewed tussock leaves.

Voice
- Silent except during breeding season, when males make a repeated 'booming' sound to attract females. This call is very similar to the 'booming' sound made by the male Bittern. It is uttered every 2 to 3 seconds and is repeated for several hours at night during the breeding season. A metallic 'chinging' call is made, apparently when females approach. Males also utter a harsh squawking sound, probably used in aggression to other males.

Food
- A wide variety of seeds, fruit, shoots and rhizomes.
- In the wild Kakapo chew leaves of snow tussock to extract juices.

- Kakapo are now managed by the Department of Conservation, and receive supplementary food.
- Fruit of the rimu tree appears to stimulate breeding.

Breeding
- Kakapo indulge in a lek breeding display which is unique in a flightless parrot. The lek or arena consists of two or more scraped hollows, or 'bowls', in the ground. These are connected by carefully manicured (trimmed) tracks. The male Kakapo boom from these bowls at night during the breeding season to attract females.
- *Time:* Breeding only occurs every third or fourth year, appears to depend on fruiting of rimu and other plants. Recently, some birds on Codfish Island have bred in consecutive years, probably due to supplementary feeding.
- *Nest:* Nest in shallow burrow, beneath tree roots or in hollow log.
- *Eggs:* A clutch of 2 to 4 white eggs are laid at intervals of 3 days. Female alone incubates eggs and feeds chicks.

- *Chicks:* As incubation commences with the laying of the first egg, the chicks hatch after 30 days and are disproportionate in size. They are fed on regurgitated vegetable material. Chicks fledge when 11 or 12 weeks old and stay with the female for several months.

Distribution & Habitat

- In 2000 the population was estimated to be 65 birds, located on Codfish and Maud Islands.
- A few males may still survive in the wilds of Fiordland.
- All known wild birds, mostly from Stewart Island, have been captured and relocated to predator-free islands.
- The population introduced to Little Barrier Island have now been removed, owing to poor breeding success.

A Kakapo eating flax seeds.

Sulphur-crested Cockatoo
Cacatua galerita

Family CACATUIDAE
Genus *Cacatua*

Category
- Introduced, probably from escaped cage birds.

Characteristics
- 50 cm.
- White with sulphur crest.
- Very wary and difficult to approach in the wild.
- Usually seen in flocks on the ground and in isolated stands of trees.

Voice
- A raucous, penetrating screech.

Food
- Fruits; seeds; buds; leaves. Also take insects, especially during nesting season.

Breeding
- *Time:* October to January
- *Nest:* Nests in tree cavities, usually high up. Nests also reported in baled haystacks.
- *Eggs:* 2 to 4 white eggs. Incubated by both parents for 30 days.

- *Chicks:* Chicks fledge when 6 weeks old.

Distribution & Habitat
- Inhabits open country where there are clumps of trees for roosting.
- Well established on west coast of North Island, from Waikato Heads to Raglan. Scattered populations in Wairarapa and Wellington Provinces.

Kookaburra
Dacelo novaeguineae

Family ALCEDINIDAE
Genus *Dacelo*

Category
- Introduced. Uncommon.

Characteristics
- 45 cm.
- A large forest kingfisher.
- Easily recognised by its overall brownish colour, white head and heavy bill.
- Often seen perched on power poles.

Voice
- Loud 'koo-koo-koo-kooah-ha-ha'. Usually heard in early morning.

Food
- Insects; lizards; mice; small birds.

Breeding
- *Time:* October to January.
- *Nest:* Nest in cavity of pohutukawa tree or tunnel bored in rotten tree.
- *Eggs:* 2 to 4 white eggs are incubated for 22 to 24 days.
- *Chicks:* Chicks sometimes do not survive, probably due to shortage of food.

Distribution & Habitat
- Introduced to Kawau Island in the Hauraki Gulf from Australia in 1860s.
- Scattered populations from Wellsford area in Northland to northern Waitakere Ranges, in open country with trees. Reports also from Banks Peninsula in the South Island.

Yellowhead / Mohua
Mohoua achrocephala

Family PACHYCEPHALIDAE
Genus *Mohoua*

Category
• Endemic. Threatened.

Characteristics
• 15 cm.
• Olive green with yellow head and chest. Female and immature: less yellow on nape.
• Confined to forests; usually seen in groups feeding in beech forest canopy.

Voice
• Males: canary-like whistle. Musical chatter in groups.

Food
• Insects, spiders and fruit.

Breeding
• Often polygamous.
• *Time:* October to December.
• *Nest:* Grass and moss, similar to Whitehead (*CB2*, 40), but in a hole or tree cavity.
• *Eggs:* 2 to 4 pinkish eggs with russet spots. Incubation entirely by female for 20 to 21 days. Sometimes double brooded, that is, they rear a second brood during the season. Fledge after 21 or 22 days. Young from previous nesting may help feed chicks.

Distribution & Habitat
• South Island only; mainly in beech forests of Eglinton Valley, Arthur's Pass and Mount Aspiring National Parks.
• Population has declined rapidly in recent years owing to predation by stoats and rats, which often kill nesting females, resulting in a disprortionate number of male birds.

Yellowhead at nest.

Male Stitchbird.

Female Stitchbird.

Family MELIPHAGIDAE
Genus *Notiomystis*

Category
• Endemic. Rare.

Characteristics
• 19 cm.
• A very active honeyeater.
• Male easily recognisable, as is unlike any other species.
• Female resembles female Bellbird (*CB2*, 48), but white wingbar and erect tail distinguish it.

Voice
• A high-pitched 't-zee'. Male also has lower-pitched warbling song, heard during nesting season.

Food
• Nectar and fruits. Insects, particularly during nesting season.

Breeding
• *Time:* Several broods a year from September to March.
• *Nest:* Nest built of twigs and fine grasses in hole in tree. Usually built above entrance hole.
• *Eggs:* Clutch of 3 or 4 white eggs are incubated by female for 15 to 17 days.
• *Chicks:* Male assists feeding chicks, which fledge when approximately 30 days old.

Distribution & Habitat
• A good population exists on Little Barrier Island since the extermination of feral cats.
• Some birds have also been relocated to predator-free islands, such as Tiritiri Matangi, Kapiti, Mokoia and Hen Islands.

Male Cirl Bunting.

Female Cirl Bunting.

Cirl Bunting
Emberiza cirlus

Family EMBERIZIDAE
Genus *Emberiza*

Category
- Introduced. Uncommon.

Characteristics
- 16 cm.
- Sparrow size.
- Greyish-brown rump distinguishes it from the common Yellowhammer (*CB1*, 39), which has a russet-coloured rump. Male has black bib.

Voice
- Song is a high-pitched buzzing rattle; rather like a cricket.
- Also a contact 'zitt' note.

Food
- Mainly seeds. Insects in the breeding season.

Breeding
- *Time:* October to January.
- *Nest:* Nest of dried grasses built in thick scrub.
- *Eggs:* Clutch of 3 or 4 greenish-grey eggs with dark streaks. Female incubates eggs for 13 days and is fed on nest by male.
- *Chicks:* Chicks fledge when 13 or 14 days old.

Distribution & Habitat
- Introduced from Britain in 1860s, where it is now rare.
- Inhabits open country with scrub, in the drier areas of Marlborough and North Canterbury.

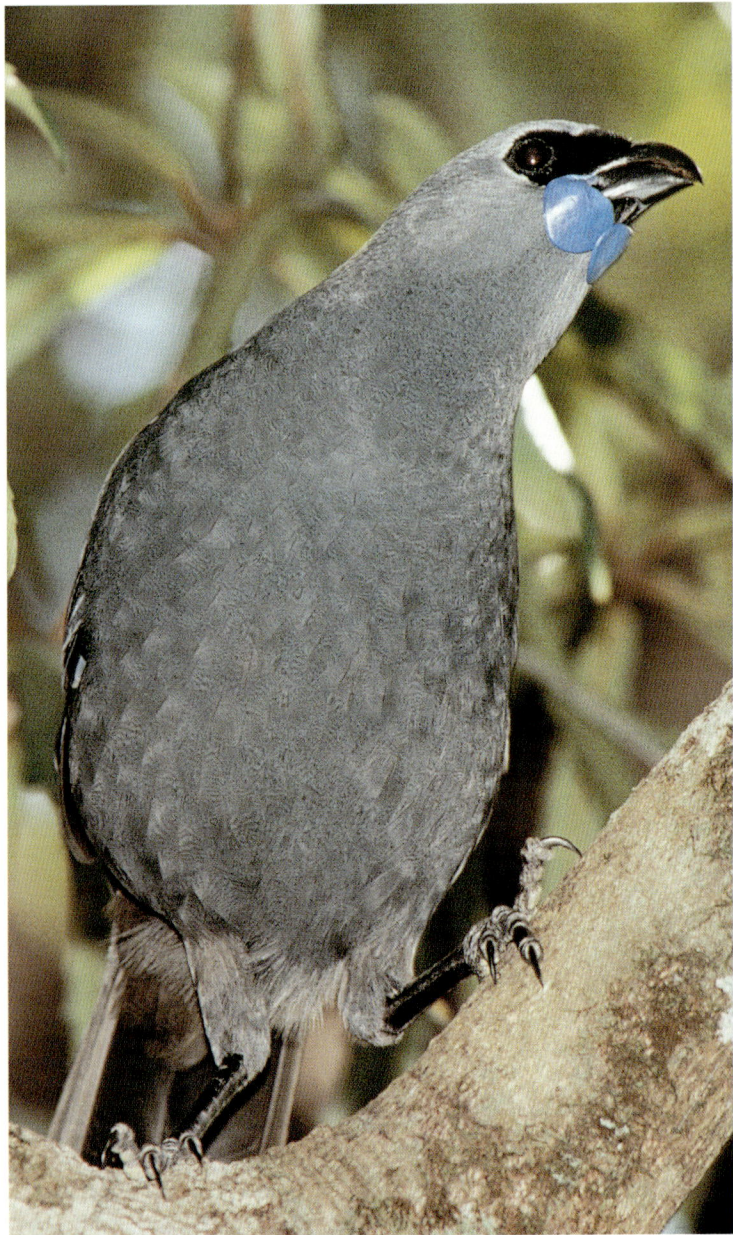

38 North Island Kokako
Callaeas cinerea wilsoni

Family CALLAEIDAE
Genus *Callaeas*

Category
- Endemic. Rare and endangered.

Characteristics
- 38 cm.
- Has distinctive blue wattles at base of a black beak.
- A weak flier, the Kokako progresses through the forest with bounds and glides.
- Skulking in habit.
- Best viewed soon after dawn, when birds sing from prominent perches.

Voice
- Melodious flute-like notes, followed by mews and clucks. Some passages resemble those of a Tui (*CB2*, 49).

Food
- Mainly vegetarian, taking a wide range of foliage, flowers and fruits.
- Also takes small insects, especially during nesting season.

Breeding
- *Time:* October to February.
- *Nest:* Bulky nest has base of sticks covered with thick bed of moss and tree fern scales. Built in shrub or dense supplejack vine. Also nests in crown of tree fern and in clumps of epiphyte.
- *Eggs:* Clutch of 2 or 3 cream eggs with brownish spots. Incubation mainly by female for 20 to 21 days.
- *Chicks:* Both parents feed chicks, which fledge when $4\frac{1}{2}$ weeks old.

Distribution & Habitat
- Inhabits native forests of Northland, Hunua Ranges, Mamaku, Rotoehu and Urewera.
- Recent successful liberations on Little Barrier, Kapiti and Tiritiri Matangi Islands.
- Intensive control of predators in the Mapara forest, King Country, has resulted in spectacular increase in breeding success.
- South Island Kokako may be extinct. Recent sightings have not been confirmed.

North Island Kokako in flight, showing small rounded wings.

North Island Kokako at nest. Note the chicks' pink wattles.

North Island Saddleback / Tieke
Philesturnus carunculatus rufusater

South Island Saddleback
Philesturnus carunculatus carunculatus

Family CALLAEIDAE
Genus *Philesturnus*

Category
• Endemic. Rare.
• South Island subspecies threatened.

Characteristics
• 25 cm.
• North Island Saddleback has narrow, buff-coloured band in front of saddle. South Island subspecies lacks this band, and its immature chicks are a uniform olive-brown, only developing the distinctive russet saddle after their first moult.
• Saddlebacks are very active, but are weak fliers.

Voice
• Usual call is a penetrating 'cheep-tee-tee-tee'. Male utters several more melodious calls.

Male North Island Saddleback.

Food

- Saddlebacks take a wide range of insects, grubs and spiders. Large insects and case moths are eaten while held with one foot, parrot fashion.
- They are vigorous feeders, prising off bark and turning over leaf litter in their search for food.
- They also eat fruits and take nectar from flax flowers.

Breeding

- *Time:* When food is plentiful Saddlebacks may have several broods a year.
- *Nest:* Nest built in hollow of tree, rock crevice or dense epiphytes, and constructed of small sticks, grasses, bark and tree fern scales.
- *Eggs:* Clutch of 2 to 4 creamy-grey eggs with blotches and dark marks on the wide end. Female incubates for 18 to 20 days and is fed by male.

- *Chicks:* Both parents feed chicks, which fledge when 3 weeks old.

Distribution & Habitat

- The North Island Saddleback is the more common subspecies. Following the first transfers from Hen Island in 1964, it has thrived on many predator-free offshore islands and on Mokoia Island in Lake Rotorua.
- Fewer than 700 South Island Saddlebacks survive on predator-free islands off Stewart Island.
- Owing to the threat of mammalian predators, and the habit of Saddlebacks to feed on the ground, the species will not be released on the mainland of New Zealand, except perhaps in predator-free mainland 'islands'.

Male North Island Saddleback, showing narrow buff band in front of saddle.

Female North Island Saddleback at nest.

The wattles of this immature North Island Saddleback have not yet developed.

Further Reading

Gill, Brian and Moon, Geoff. *New Zealand's Unique Birds*. Reed Books, Auckland, 1999.

Heather, B.D. and Robertson, H.A. *The Field Guide to the Birds of New Zealand*. Penguin, Auckland, 1996.

Moon, Geoff. *The Reed Field Guide to New Zealand Birds*. Revised edition. Reed Books, Auckland, 1998.

Index of Common and Maori Names

Index of Scientific Names